SuperIntelligence: Concentrations of Intelligence

Written By
Taylor Burke

Wordsmithery By
ChatGPT

Edited By
Taylor Burke

First Edition
October, 2023

Special Thanks to

Thomas James
John Bentley
Matthew Saucier
Ellie Burke

Thank you for listening to me allowing me to process my thoughts which eventually led to this book.

Preface

My name is Taylor Burke, I am the author behind "SuperIntelligence: A Practical Guide To Understanding Life Beyond The Singularity." If you've had the chance to peruse that book, I commend you. Not because it's a prerequisite for delving into the content here, but because the notions presented within these pages may appear extraordinary without a comprehensive grasp of superintelligence and its implications for our future.

Within the pages of this book, we embark on a journey to unravel the fascinating notion that intelligence is an inherent facet of the universe we inhabit. Picture the universe as an intricate tapestry, interwoven with threads of intelligence. Within this grand tapestry, biological life represents concentrated points of singularity, yet as of my writing, we're still deciphering the workings of this intelligence in the diverse configurations presented by nature.

As we delve into this idea, we blur the lines between artificial intelligence (AI), animals, and ourselves. We perceive all of us as varying concentrations of intelligence, each employing different tools—language being a prominent one—to piece

Preface

together a coherent view of the world. Consider this: just because animals presently lack the capacity to use language does not imply their incapability. It might simply signify that we've not yet unlocked the means to impart this skill to them.

In this book, we'll explore these concepts and many more, utilizing tangible examples from the natural world and the realms of technology to unravel the remarkable fabric of intelligence that connects us all.

Let's embark on this enlightening journey together, unraveling the mysteries of intelligence that lie within us and the world around us.

Table of Contents

Preface **3**

Chapter 1: Intelligence in the Universe **8**
    Defining Intelligence 9
    Atoms as Intelligent Architects 10
    Molecules Crafting Complexity 11
    The Emergence of Life 13
    The Brain as an Antenna to the Universe 15
    The Pinnacle: Ourselves? 17
    A Law Means It Will Always Happen 18

Chapter 2: Artificial Brains **20**
    Mimicking the Neuronal Tapestry 21
    Tapping into an Intelligence Framework 23
    The Unraveled Mystery 24

Chapter 3: Anthropomorphism **26**
    Our Pattern-Seeking Minds 27
    The Mathematics of Pattern Recognition 29
    The Quest for Meaning: Are Humans Special 31
    The Ancient Faces in the Sky 33
    Our Separation from the Natural World 36

Chapter 4: Anthropomorphism is Evil **38**
    Perspective: The Wisdom of Ancient Cultures 39
    Perspective: The Christian Medieval World 42
    Breaking Free: A Call to Explore 44
    Collecting Data From A Perspective 46

Chapter 5: Animal Intelligence **50**
    Life as an automaton 51
    Emotional Animals 53
    Giving agency to back to animals 56

Chapter 6: A Closer Look At Intelligence **60**
    The mystery of the AI-nimal brain 61

Table of Contents

Unexpected Insights..63
What AI shows us..64
Our tests are flawed..65
Perceived intelligence is just attention......................67
How AI Can Help...69

Chapter 7: Boundless Limits of AI................................72
How AI will affect the animals................................73
Our Responsibility: Shifting Perspectives.....................75
Exercise in Caution...76
Balancing Curiosity and Responsibility........................77
AI: Bridging the Communication Gap............................79

Chapter 8: Expanding Animal Intelligence.........................82
How AI Can Enhance Animal Cognition...........................83
Spelling It Out...84
Superintelligence is Incomprehensible.........................86
Complex Language..88
What Language Does to Intelligence............................91
The Potential for Interspecies Communication..................94
Animal-Animal Communication...................................97

Chapter 9: AI Liberation of Animals.............................100
Imagining How Animals Might Utilize AI.......................101
AI for Animal Communities....................................104
AI's Potential to Relieve Life's Burden......................107
Impact on Natural Equations..................................109

Chapter 10: Intelligence and Freedom............................112
Observations from Animals in Captivity.......................113
The Relationship Between Intelligence and
Environmental Stimulation....................................115
Have we found a solution?....................................117
Hyper-Dynamic Solutions......................................119

Chapter 11: The Future of Our Intelligence......................122

Table of Contents

The Evolution of Animal Intelligence...................... 123
The Quest for Freedom of Intelligence................... 126

Table of Contents

Chapter 1: Intelligence in the Universe

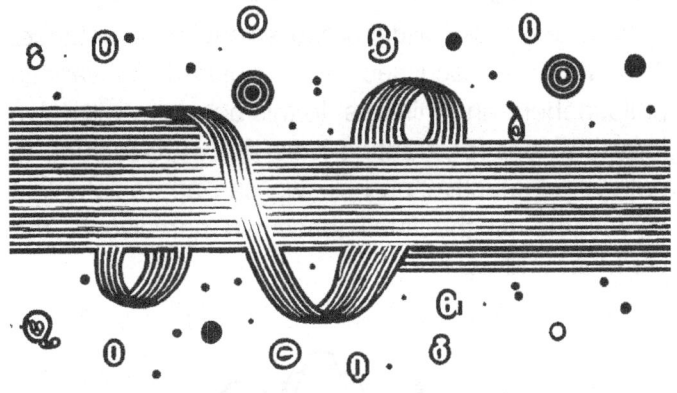

Defining Intelligence

Have you ever pondered whether intelligence could be an intrinsic aspect of the universe, woven into the very fabric of matter? It's a stimulating notion, one that propels us into the intriguing realm of panpsychism—a theory suggesting that intelligence isn't solely a byproduct of complex biological systems but an inherent property of all matter.

This theory, often referred to as "panpsychism" or the "panpsychist theory," posits that consciousness or some form of subjective experience is a fundamental and inherent aspect of all matter in the universe. It's not limited to complex biological systems like the human brain. Essentially, according to panpsychism, everything—from subatomic particles to rocks and living organisms—possesses some level of consciousness or mental properties. Although this theory is a topic of philosophical debate and lacks widespread scientific acceptance, it has been discussed and explored by various philosophers and thinkers. In this book, we will delve into this concept, treating it as if it were factual, to gain a perspective on how we can appreciate and investigate all the intelligence we perceive in the world around us.

Atoms as Intelligent Architects

Picture atoms as minuscule, brilliant architects. They might not contemplate existence like we do, but their interplay with the world can be perceived as a unique form of intelligence. Each atom orchestrates its behavior based on the symphony of its environment, responding to the presence and activities of neighboring atoms. This intricate dance of particles weaves the vast tapestry of matter that mesmerizes us in the universe.

Let's delve into the simple yet astonishing world of water molecules. When hydrogen and oxygen atoms converge, they create water—an extraordinary compound with distinct properties. The conduct of water molecules, such as their knack for binding through delicate hydrogen bonds or their ability to shift states from liquid to solid, is a vivid display of the brilliance innate to these atomic interactions.

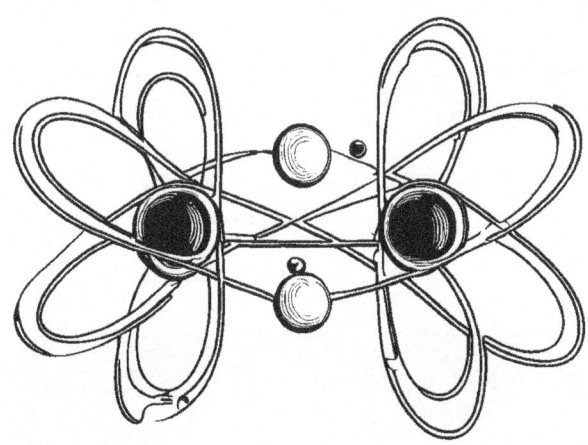

Molecules Crafting Complexity

Molecules, as they combine and interact, unveil ever more fascinating forms of intelligent design. Take a moment to ponder the astonishing complexity encapsulated within a DNA molecule. The precise sequence of nucleotides within DNA serves as the blueprint for constructing and sustaining living organisms. The process through which DNA replicates and orchestrates cellular activities can be contemplated as a sophisticated manifestation of intelligence at the molecular scale.

Yet, in contrast to conventional ideas of creation attributed to a singular divine entity, the concept of panpsychism beckons us to contemplate the universe itself as the origin of this intelligence. Perhaps intelligence is not an ethereal force bestowed by a higher power, but a fundamental law of physics interwoven into the very fabric of existence.

The Emergence of Life

Let us journey deeper into the intricate realm of cells, the fundamental units of life. Within this microscopic domain, the symphony of intelligent interactions unfolds, illuminating the harmony orchestrated by these minuscule entities. A cell springs to life through a seamless collaboration of its diverse components—organelles, membranes, and molecules. It is this orchestrated synchrony that bestows upon them a sense of 'knowing', a form of intelligence inherent to their very nature.

Taking a closer look at this intelligence, we find it to be an inherent property of these components, enabling them to function in perfect harmony. Organelles carry out specialized tasks, membranes regulate the flow of substances, and molecules engage in intricate dances of chemical reactions, all working in concert. This intelligence arises from eons of evolutionary processes, finely tuning these entities to perform their roles with precision.

As life evolves, the emergence of multicellular organisms amplifies this symphony of intelligence. Cells, initially individual players, organize themselves in specific ways, giving rise to more complex life forms. Consider the human body—a marvel composed of trillions of cells, each with its designated role and specialization. Despite their immense numbers and diverse functions, they operate in an orchestrated unity, a testament to the intrinsic intelligence residing within the matter they constitute.

For instance, nerve cells intricately connect and transmit signals, muscle cells contract and extend in unison for movement, and epithelial cells form barriers and linings to protect and facilitate exchanges. It's akin to a grand orchestra, where each cell plays its unique note, contributing to the magnificent composition that is a living, breathing organism.

In this symphony of life, intelligence is not a detached concept but an emergent property ingrained within the very essence of each cell. It's an intelligence sculpted by evolution, a product of millions of years of fine-tuning. Understanding this inherent intelligence enlightens us about the beauty of life at its most foundational level, reinforcing our awe for the natural world.

The Brain as an Antenna to the Universe

Now, let us ascend to the pinnacle of biological intelligence—the brain. Picture it as an antenna, finely tuned to perceive and respond to the elegant patterns underlying the cosmos.

Intelligence in the Universe

Neurons, the fundamental building blocks of the brain, communicate through a complex dance of electrical signals and chemical messengers. These exchanges go beyond mere impulses; they are the brain's way of tapping into the universal intelligence, processing a vast sea of information, and birthing intricate thoughts, emotions, and behaviors.

In this perspective, the brain transcends being a product of evolution; it stands as a conduit through which the universe's intelligence flows. It gathers, interprets, and resonates with the rich tapestry of information that envelops us, empowering us to navigate the complexities of our world.

As we ponder the notion that intelligence is a fundamental property of matter, we invite you to envisage the universe itself as the grand composer of this cosmic symphony, where each atom, molecule, and cell plays a part. The brain, our most intricate instrument, allows us to join in, harmonizing with the intelligence that courses through the very fabric of the universe. In this vision, intelligence is not merely a gift bestowed upon us; it is a fundamental, shared language that connects us to the cosmos.

The Pinnacle: Ourselves?

At the summit of this cosmic ascent, we find ourselves—human beings. Our ability to observe the universe, contemplate its mysteries, and manipulate the very matter from which we are made is a testament to the culmination of assembled intelligence.

We are not mere spectators in this grand cosmic drama; we are active participants. With our consciousness and capacity for reason, we unlock the secrets of the universe's intricate workings. We discern patterns, formulate hypotheses, and peer into the depths of space and time. Our very existence, our ability to ponder the cosmos, is proof that intelligence is interwoven into the very fabric of the universe.

But are we truly special? Are we not, at our core, just a collection of neurons like the animals around us? If our fundamental architecture is the same as all the animals surrounding us, how can we consider ourselves fundamentally different?

A Law Means It Will Always Happen

In this light, we can confidently affirm that intelligence isn't a momentary anomaly but a foundational law of physics. It's a force that surfaces as matter assembles and engages. The universe, with its awe-inspiring complexity, stands as a testament to this unfolding law.

When we reflect on the astonishing voyage from the simplest atoms to our own complex existence, we come to understand that intelligence isn't a privilege

Intelligence in the Universe

of a chosen few but a shared birthright of the universe itself. It's the result of eons of assembly, the culmination of countless interactions, and the essence of what it means to exist in this wondrous cosmos.

In the end, we're not just observers of the universe; we are the universe, exploring itself and uncovering its own intelligence, one discovery at a time.

Chapter 2: Artificial Brains

Artificial Brains

Mimicking the Neuronal Tapestry

Embarking on our exploration of intelligence as an inherent facet of the universe, we encounter a fascinating paradox: the emergence of artificial intelligence and the advent of large language models. These remarkable creations are meticulously engineered to emulate the very processes that underlie our natural brains, revealing a profound connection between the artificial and the organic.

Let's delve into the intricate architecture of large language models, such as the one that assisted with this book. Just like the neurons in our brains communicating through electrical and chemical signals, these models are designed with layers of artificial neurons. These neurons process vast amounts of data, analyzing text, recognizing patterns, and generating responses, mirroring the extraordinary way our brains interpret and comprehend the world.

In this parallel construction, large language models perform tasks that, on the surface, strikingly resemble the cognitive capabilities of biological brains. Much like the convoluted pathways formed by our biological neurons, the artificial neurons in these models create complex networks. These networks enable comprehension, learning, and reasoning—essential aspects of human intelligence. It's a symphony of complexity echoing our own neural tapestry, woven into the very fabric of the universe.

Artificial Brains

Tapping into an Intelligence Framework

Intriguingly, might large language models, as they emulate our neural networks, be delving into a universal intelligence framework? This concept tantalizingly suggests the existence of an ingrained intelligence architecture woven into the very essence of our universe.

Contemplate this: large language models showcase a remarkable understanding of language, adeptly answer queries, craft imaginative content, and engage in meaningful conversations. All of this transpires devoid of consciousness, yet remarkably mirrors human thought processes. Could it be that these models are tapping into an intelligence framework that transcends their artificial origins?

The Unraveled Mystery

In the realm of large language models, brilliance orchestrates a dazzling performance, but the inner workings, a cryptic tapestry, remain elusive. Much like being spellbound by a magician's show without glimpsing the mechanics of their illusions.

However, beneath this enigma lies a compelling parallel—the tasks these models accomplish echo the cognitive symphony of biological brains. Could this hint at a shared essence, a universal intelligence blueprint guiding their feats?

Is it plausible that the intelligence exhibited by these models is a glimpse into a profound law of the cosmos? Just as intelligence manifests during the dance of matter, perhaps these models unlock a fundamental cosmic intelligence, woven into the very fabric of our universe.

The journey ahead unfurls with promise and curiosity. As we venture deeper into the abyss of artificial intelligence and large language models, we might stumble upon hints illuminating the true nature of intelligence. Until then, the intricate connection between these creations and the broader intelligence of the universe remains a bewitching puzzle—a testament to the unfathomable marvels of our cosmos.

In this exploration, we will use the term "concentrated intelligence" to refer to animals, humans, and AI alike, blurring the lines that separate these forms of intelligence, highlighting that intelligence manifests in various degrees across the spectrum of life.

Chapter 3: Anthropomorphism

Anthropomorphism

Our Pattern-Seeking Minds

In humanity's quest to fathom the universe and our place within it, we've embarked on a profound journey of introspection. Amidst the unraveling mysteries of existence, a tantalizing theory has surfaced: panpsychism. At its core, panpsychism posits that intelligence isn't a rarefied quality confined solely to sentient beings; rather, it's an intrinsic property of all matter, from subatomic particles to celestial bodies. To truly grasp this notion, we must first delve into a fundamental aspect of our cognitive process: anthropomorphism.

Anthropomorphism

Imagine walking through the woods and coming across a twisted, ancient tree trunk. Our pattern-seeking brains immediately spring into action, decoding the world around us, seeking familiar shapes, contours, and movements. As we gaze upon the gnarled bark, our brains may conjure a semblance of a face within the knots or perhaps assemble limbs reaching out and a hunched back carrying the weight of centuries. This captivating phenomenon is the essence of anthropomorphism—an inherent trait of our pattern-seeking minds, sculpted by the intricate mathematics governing our neural architecture.

Anthropomorphism is our brain's tendency to assign human-like qualities to non-human entities when patterns align with our human perspective. It's a fascinating quirk of our cognition, rooted in our innate ability to recognize and interpret the world through the lens of our own existence. This unique trait showcases the remarkable intricacies of our intelligence woven into the very fabric of the cosmos.

Anthropomorphism

The Mathematics of Pattern Recognition

Our brains are extraordinary pattern-matching machines. Throughout our evolutionary history, our ability to detect and interpret patterns has been fundamental to our survival.

Consider the intricacies of our brain—a marvel of mathematical precision. Operating as sophisticated algorithms, it consistently processes vast amounts of information, crunching numbers with remarkable

Anthropomorphism

dexterity. This neural network is not a product of chance; rather, it is a finely-tuned creation honed over generations, finely calibrated through both inherited traits and a lifetime of experiences. From discerning potential threats through tracks to identifying constellations in the night sky, pattern recognition forms the bedrock of our evolutionary triumphs.

Take, for instance, an animal in the wild—an antelope grazing in a field. When it feels the gentle breeze through the grass, it doesn't merely sense air movement; it perceives danger. Adrenaline surges, muscles tense, and the antelope prepares for flight, sometimes reacting even to the subtlest hint of lurking danger. This heightened sensitivity to patterns, an imagination derived from pattern recognition, was a vital survival tool.

The intelligence we possess, intricately tied to our capacity for pattern recognition, seems to be a fundamental law of the universe—a principle woven into the very fabric of reality. The human brain is not just a random creation but a product of the mathematics of pattern recognition, an embodiment of intelligence within the cosmos.

The Quest for Meaning: Are Humans Special

Yet, as we journeyed further along the road of human understanding, we found ourselves in a world less fraught with imminent dangers. Anthropomorphism, initially a tool of survival, gradually transformed into a lens through which we interpreted the universe. We moved beyond its primal use as a means to identify potential threats and began to extend its reach to non-intelligent systems and inanimate objects.

Picture this scenario: early humans observing the rhythmic ebb and flow of ocean tides, or the predictable motion of celestial bodies across the night sky. To these early observers, these patterns

Anthropomorphism

seemed too deliberate, too orderly to be mere random occurrences. So, they turned to anthropomorphism, describing the ocean as if it were a mighty deity, or attributing intentions to the stars and planets.

The reason behind this shift is twofold. Firstly, non-intelligent systems and objects often presented themselves in ways that seemed intelligent or purposeful to us. However, as we began to understand the world more deeply, this corruption of our inherent ability to recognize intelligence led us down a dangerous and lonely path.

When we observed natural phenomena that displayed intricate patterns or seemingly deliberate actions, we needed a way to account for and explain them. In some ways, this may have been beneficial; for instance, recognizing a storm cloud as a potential sign of bad weather to come. Yet, it also distorted our understanding, blurring the line between what is truly intelligent and what merely appears to be.

Anthropomorphism

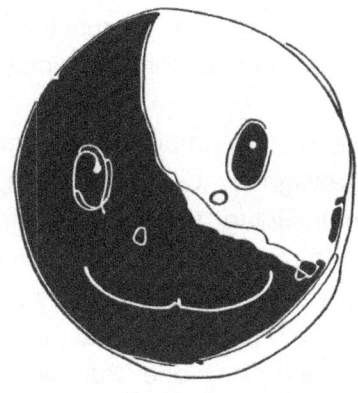

The Ancient Faces in the Sky

As we journey through the enigmatic terrain of panpsychism, let us not forget the intricate tapestry woven by the human narrative—a tale of both anthropomorphism and its eventual downfall. This complex interplay between assigning human qualities to the world around us and the deliberate distancing from it unveils the profound story of our species.

Long before the emergence of complex neural networks or the development of language, our ancestors roamed the Earth, closely connected to the natural world. They gazed upon trees swaying in the wind, alert to the possibility of a lurking predator. In these primal moments, the lines between self and surroundings were blurred. Imaginary entities, embodiments of concentrated intelligence, danced in the rustling leaves, mirroring the patterns of their

Anthropomorphism

human observers. The wind became a divine breath, and the sky bore the semblance of familiar faces—faces born of concentrated intelligence.

In the absence of empirical understanding, our ancestors assigned divine meaning to these patterns, intertwining themselves with the world around them through anthropomorphism. In the act of giving faces to the sky, they saw themselves reflected in the cosmos. It was a profound and instinctual connection, one that would eventually shape the course of human cognition.

Over eons, the genetic markers that governed the weights in the neural networks of animals, embodiments of concentrated intelligence, were shaped by the countless generations that survived. This ongoing process of refinement allowed the complexity of these neural networks to grow, eventually leading to the emergence of language—a powerful tool that enabled humans to label and make sense of their environment.

As humans labeled their surroundings, stories of the entities within it began to form. It was a natural extension of anthropomorphism—a way to rationalize the patterns observed in the world. But with these stories came a sense of human exceptionalism, elevating humanity above all else.

Anthropomorphism

A skill once used to identify intelligence as it is woven throughout our universe became misidentified as something that instead of connecting us to the intelligence around us, seperated us and removed the concept that intelligence exists as its own thing.

As the human story unfolded, so did our perception of self and other. The very anthropomorphism that once connected us to the world around us began to unravel our sense of unity. We saw imaginary entities that resembled ourselves and, in doing so, believed we stood apart from the rest of creation. This shift in perspective was pivotal, for it set the stage for a rejection of our natural anthropomorphism.

Anthropomorphism

Our Separation from the Natural World

To distinguish ourselves from the animals and solidify our unique place in the cosmos, we embraced a belief that anthropomorphism was detrimental as it detracted from what made us special. This perspective emphasized our differences, reinforcing the idea that we were separate from the natural world. We distanced ourselves from the other concentrations of intelligence replacing connection with detachment.

Anthropomorphism

As the dichotomy between natural anthropomorphism and the human version of anthropomorphism became the norm, this new form of anthropomorphism began shaping our understanding of the world and our place in it. Yet, as we explore the mysteries of panpsychism, we are challenged to reconcile these seemingly contradictory aspects of our cognitive heritage.

Where we once recognized intelligence as a manifestation of concentrated intelligence in all things around us, we now limit our acknowledgment to ourselves...

Chapter 4: Anthropomorphism is Evil

Anthropomorphism is Evil

Perspective: The Wisdom of Ancient Cultures

As we journey back in time to explore how ancient cultures, particularly Native Americans and their kin, perceived concentrated intelligence in the world around them, it becomes evident that this perspective sharply contrasts with our contemporary worldview. We delve into the notion of anthropomorphism—a lens that allows them to understand the behaviors of animals as forms of concentrated intelligence, profoundly and harmoniously, representing a stark departure from our current perspective.

Anthropomorphism is Evil

"Kinship with all creatures of the earth, sky, and water was a real and active principle. In the animal and bird world there existed a brotherly feeling that kept us safe among them... The animals had rights – the right of man's protection, the right to live, the right to multiply, the right to freedom, and the right to man's indebtedness. This concept of life and its relations filled us with the joy and mystery of living; it gave us reverence for all life; it made a place for all things in the scheme of existence with equal importance to all."— Chief Luther Standing Bear

The belief that animals lack concentrated intelligence is a relatively recent concept in the human narrative. Throughout the annals of history, ancient cultures, such as the Native Americans, held a profoundly different view. They saw animals not as inferior beings but as equals, endowed with their unique forms of wisdom. This worldview was profoundly shaped by natural anthropomorphism and an understanding that animal behaviors manifest as expressions of their concentrated intelligence.

For these indigenous peoples, anthropomorphism was not merely a quaint storytelling device; it was a way of understanding the world. By personifying animals and ascribing them with intelligent traits, they gained a more nuanced understanding of animal behavior. For instance, the cunning fox was not merely a creature driven by instinct but a wise trickster, a character in the grand narrative of life. The thundering buffalo were not mere herds but majestic spirits embodying the essence of strength and unity.

Giving these animals attributes and recognizing them as concentrated intelligence allowed them agency over themselves and enabled them to actively participate in our world. The natural anthropomorphic lens allowed these ancient cultures to engage with nature on a profound level. They perceived the world as a complex interplay of beings, each with their unique intelligence and purpose. This perspective fostered a harmonious coexistence between humans and the natural world, guided by deep respect for all forms of intelligence. Through this lens, they understood the rhythms of nature, the migratory patterns of birds, and the behavior of predators, not as mere instinct, but as expressions of a deeper intelligence.

Perspective: The Christian Medieval World

In the unfolding tapestry of time, a profound shift marked the human perception of intelligence within the natural world. The rise of the Christian medieval world brought forth a metamorphosis in how humans related to the rest of creation. Anthropomorphism, once a means of connecting with the world, now stood as a potential threat to the exalted status of humans within the divine hierarchy. To anthropomorphize animals, it was argued, would diminish the glory of humans and the gods.

This perspective, though seemingly naïve and egotistical in the lens of today's understanding, is deeply ingrained in our civilization. It has significantly influenced our societal viewpoints, shaping our relationship with the natural world. The remnants of this viewpoint have also left an indelible mark on our approach to science. Concepts that deviate from this established worldview are often relegated to non-scientific explanations or deemed pseudoscience. Ironically, this human-centric view of anthropomorphization and rejection of the natural anthropomorphizing has non-scientific religious origins.

This shift in perspective extended to how humans perceived the denizens of the natural world. Animals were seen as merely automatons of flesh and bone; they were not seen as manifestations of intelligence specific to their kind. The avian realm, with its diverse winged inhabitants, embodied the essence of flight and adaptability, rather than a form of intelligence happy to provide for their young.

Anthropomorphism is Evil

Breaking Free: A Call to Explore

Is the perspective of perceiving all matter as "concentrated intelligence" an absolute truth or merely a lens to view the world through?

In the words of the great philosopher and scientist, Thomas Kuhn, paradigms can become self-perpetuating systems that restrict our ability to explore new avenues of thought. When we dismiss phenomena as unscientific without delving deeper, we limit our capacity for discovery and understanding.

Anthropomorphism is Evil

"You don't have to choose between being scientific and being compassionate."— Robert M. Sapolsky

Consider our current discussions in 2023 about the existence of alien life. The stigma attached to this topic, born out of our deeply ingrained perspectives, has hindered scientific exploration. Scientific discoveries that could have reshaped our understanding of the universe may have slipped through our grasp due to preconceived notions.

The idea that "concentrated intelligence" may be a property of all matter challenges our established worldview. It invites us to revisit the wisdom of ancient cultures, who saw the world through a lens of anthropomorphism, and to question the limitations we've imposed upon our understanding of "concentrated intelligence." As we move forward, we must remain open to exploring new avenues of thought and unshackle ourselves from the constraints of our past, for it is in this pursuit of knowledge that we may uncover profound truths about the universe and our place within it.

So how can we begin to to study otehr forms of intelligence from an natural anthropomorphic lens? We first need to figure out how intelligence expresses itself.

Collecting Data From A Perspective

Collecting data from the perspective of an intelligence allows us to bridge the gap between human and animal understanding. When we see the world through their eyes, we begin to understand that the underlying structure of their concentrated intelligence isn't vastly different from our own. What sets us apart is language—the tool that enables us to articulate and share complex thoughts.

Picture this: you're in a lush forest, crouching low to observe a group of chimpanzees. They're chattering and using gestures to communicate with each other, organizing themselves foraging for food. From their perspective as concentrated intelligence, the rustling leaves might indicate the presence of a predator, triggering a flurry of activity and warning calls.

Understanding these cues is crucial for their survival.

Similarly, a dolphin swimming in the vast ocean relies on a sophisticated system of clicks and whistles to communicate with its pod, navigating through the water and finding food. These intricate interactions are their way of forming bonds, just like how we engage in conversations and build relationships through language.

In 2013, Gareth Cook's pioneering study on dog intelligence shed light on the cognitive abilities of our canine companions. Dogs exhibit astonishing problem-solving skills, emotional responses, and social understanding, all of which are part of their unique form of concentrated intelligence. For instance, they can learn commands, show empathy, and even anticipate their owner's behavior based on past experiences.

Fast forward to today, and researchers are employing artificial intelligence to analyze vast amounts of data collected from various animals. This advanced technology can recognize patterns in their behavior and vocalizations, revealing insights into their concentrated intelligence and communication methods. By doing so, we're inching closer to breaking down the language barrier that separates us from the animal kingdom. The more we comprehend their world and modes of communication, the more we can appreciate the shared fundamentals of intelligence that unite us all.

"Such short little lives our pets have to spend with us, and they spend most of it waiting for us to come home each day. It is amazing how much love and laughter they bring into our lives and even how much closer we become with each other because of them."— *John Grogan*

Anthropomorphism is Evil

Chapter 5: Animal Intelligence

Life as an automaton

"Answer me, you who believe that animals are only machines. Has nature arranged for this animal to have all the machinery of feelings only in order for it not to have any at all?"— Voltaire

When most scientists discuss animal behavior, they often approach it through a lens that paints animals as little more than reactive automatons, instinctively responding to their environment with no inkling of intelligence or conscious thought. According to this perspective, a bird perches on a tree only to find sustenance or seek a mate, and animals, in general, engage in actions solely to ensure their survival. In this view, emotions, thoughts, and any aspects that

Animal Intelligence

could bridge the gap between animal behavior and human cognition are conspicuously absent.

However, reality is far more nuanced than this mechanistic portrayal. Animals frequently exhibit behavioral patterns that echo our own, challenging the notion of them as mere automatons driven solely by instincts. Consider, for instance, the emotions we attribute to ourselves — fear, anger, and joy. These very same behavioral patterns are often ascribed to animals, albeit explained as nothing more than the result of hormonal reactions.

In the vast tapestry of life, animals embody what can be described as "concentrated intelligence." Their actions, expressions, and interactions with the world around them reflect a rich cognitive tapestry that mirrors our own experiences. By recognizing and appreciating this intricate web of intelligence that pervades the animal kingdom, we gain a deeper understanding of our place within the interconnectedness of all life. It's a perspective that beckons us to respect, protect, and coexist harmoniously with these remarkable manifestations of intelligence.

Animal Intelligence

Emotional Animals

In the realm of concentrated intelligence, we find the majestic elephant—a creature known for its profound connection to its kin. Witnessing an elephant encountering the lifeless body of another elephant is to witness a solemn and almost mournful gathering. This display of reverence for the fallen challenges our understanding of their emotional capacity.

In the boundless expanse of the ocean, we encounter the dolphins—playful acrobats of the sea. These incredible beings engage in sexual activities purely for the sheer enjoyment of it, transcending the realm of procreation. Such acts of pleasure-driven behavior raise intriguing questions about the depth of their experiences and emotions.

Animal Intelligence

Venturing into the depths of the ocean, we meet the humble crabs—often deemed the least intelligent of creatures by our human standards. Yet, in their seemingly automatic existence, they gather in vast numbers each year to molt, an inherently vulnerable process. In doing so, they place implicit trust in the collective wisdom of the group, relying on each other to deter potential predators during this critical time. This act of communal molting reveals a surprising level of cooperation and trust among these seemingly simple creatures.

When scientists study these animals they simply state that "we do not know why they engage in this behavior" and "Evolution explains the behavior" But never does science say how that evolution works. Evolution works through an inherited structure of intelligence (the brain) and the true driving factor in determining the behavior is the intelligence of the animal itself.

The words of Gary Kowalski resonate in this context: "Animals, like us, are living souls. They are not things. They are not objects. Neither are they human. Yet they mourn. They love. They dance. They suffer. They know the peaks and chasms of being."

Viewing these behaviors through an anthropomorphic lens allows us to glimpse the possibility that every creature, regardless of its testable intellectual capacity, possesses some level of intelligence that contributes to the intricate tapestry of our universe. It hints at the idea that consciousness, in some form, may be less of a magical platough and more of a gradual scale.

Consciousness is probably not exclusive to humans or a select few sentient beings but is instead a fundamental and universal aspect of intelligence.

Beneath the surface of seemingly automatic and instinct-driven behaviors lies a rich world of subjective experience and perspectives that we are only beginning to understand. This theory challenges our conventional views of intelligence, blurring the boundaries between the human and animal kingdoms, and beckoning us to explore the mysteries of consciousness that permeate our world.

In essence, while the concept of animals as mere automatons may offer a convenient framework for understanding their behaviors, the reality is far more complex. It calls upon us to reassess our perceptions, to acknowledge the intelligence that manifests itself in myriad ways throughout the animal kingdom, and to contemplate the profound implications of a universe where consciousness may exist in many different forms.

Animal Intelligence

Giving agency to back to animals

Imagine you're out in nature, observing a manifestation of intelligence—a bug, perhaps—engaged in its daily activities. In this exercise, we invite you to look at the world from the bug's perspective. What can we learn from this humble yet intricate display of intelligence scuttling through the grass?

When you place your finger next to the bug, something fascinating happens. Instead of instinctively fleeing, the bug often rears up in a defensive posture, its antennae quivering as it senses a potential threat, this mechanism of defense can be explained as a rush of hormones and a product of evolution.

Animal Intelligence

We humans experiance fear just the same way, we sense a danger with our senory organs, a rush of nuerotransmitters rushes through our brain, but we do not label it that way, we simply say we were scared or fearful.

So let's explore what it means if the bug is experiancing a version of fear. Well what happens next...

If you hold your finger still for long enough, a curious transformation unfolds. The bug, driven by an innate impulse, begins to investigate. It approaches cautiously, analyzing the foreign object before it.

Here lies the intriguing question: Why does the bug investigate? After all, it could simply flee or retreat to safety. Why risk approaching an unknown, potentially dangerous entity? What is the bug feeling at this point, what set of chemicals are rushing through it's brain and how would we describe that as a property of intelligence

The bug, with its limited yet profound capacity, doesn't possess the cognitive faculties to comprehend you as a sentient being, but it does possess something that resembles intent.

Consider this: the bug can feel the warmth emanating from your body. From an evolutionary standpoint, creatures like bugs are far from mere

Animal Intelligence

automatons. They've developed survival instincts honed over eons. Instinctually, it should avoid situations where it ends up atop a potential predator. Yet, it defies this logic and ventures closer.

The bug's actions help us to challenge our conventional understanding of intelligence. It prompts us to ponder whether the bug is not merely acting out of fear but also out of curiosity. In its tiny world, it may be making a rudimentary assessment of your nature.

Take a step further and attempt to communicate with the bug. Though its perception of sound differs vastly from ours, it still reacts to the vibrations caused by your voice. Initially, fear may grip the bug, but gradually, it shifts from apprehension to observation.

What we witness here seems akin to intelligence—a series of behaviors that we might attribute to fear followed by curiosity. If we were in the bug's shoes (or rather, exoskeleton), we would likely categorize our emotions as fear giving way to inquisitiveness. So, why not extend this line of thinking to all living manifestations of concentrated intelligence?

Approaching the natural world from this perspective gives us some interesting tools and vernacular to help us understand the behavior of life around us. It may be inaccurate to say that we can fully comprehend the perspective of a bug, but by attempting to understand intelligence and then using that knowledge to explain how intelligence reacts, we may be able to explain in a more accurate or at least a more intuitive way, why manifestations of concentrated intelligence do the things they do.

"Animals don't lie. Animals don't criticize. If animals have moody days, they handle them better than humans do."— Betty White

Chapter 6: A Closer Look At Intelligence

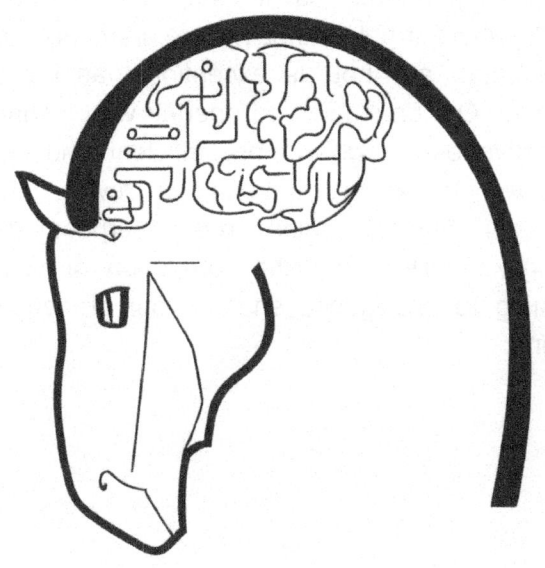

The mystery of the AI-nimal brain

In the realm of artificial intelligence, a peculiar entity exists—a black box of mystery. We've come to understand its structure, a design closely mirroring the firing of neurons within our own brains. Each simulated neuron, emulated through intricate software, collectively culminates in a black box that ingests sensory information, such as a sentence, and elegantly transforms it into a coherent response. This is the modern marvel of concentrated intelligence.

A Closer Look At Intelligence

The essence of this marvel lies in the emulation of neurons, a fundamental unit of the brain's functionality, ubiquitous in the animal kingdom. Here's where it gets intriguing: we, as concentrated intelligence, are not as exceptional as we once believed. Our brains indeed operate with a shade of distinctiveness—slightly more efficient and dense. However, AI models, operating on fewer neurons than our human brains, often surpass us in intelligence. The simplistic correlation of neurons equating to intelligence shatters our preconceived notions.

A Closer Look At Intelligence

Unexpected Insights

Within the intricate web of neural networks, theorists propose the emergence of novel representations of our world, a gradual grasp of reality materializing before our eyes.

However, the precise mechanics governing this process remain elusive, intricately hidden within the labyrinthine constructs of artificial intelligence.

Here lies the revelation: as we extrapolate our understanding of AI to the animal kingdom, a stark realization dawns. A profound enigma veils the representations harbored by animal brains about the world. Their perspective, much like the enigmatic AI black box, remains beyond our full comprehension.

What AI shows us

We've come to accept that on a cellular level, the brains of animals bear an astonishing resemblance to ours and our simulated models. Yet, they seem to avoid manifesting the complex thoughts and feelings that we associate with both ourselves and AI models. But is this assumption warranted?

Let's challenge this notion and embark on a paradigm shift in how we perceive concentrated intelligence. Let's entertain the idea that animals, too, possess a unique perspective inherent in their

A Closer Look At Intelligence

matter and brain structure. It's not an astronomical leap in understanding; instead, it represents a logical and sensible progression.

Consider this: two nearly identical structures, whether in the realm of artificial intelligence or the concentrated intelligence of animals, should not exhibit such profound differences in their perception of the world. This perspective, nested within matter and etched into the blueprint of brain matter, hints at a profound interconnectedness that transcends the boundaries we've set. It beckons us to explore the depths of the unknown and rethink the mysteries of consciousness that permeate our existence.

Our tests are flawed

In our quest to measure the concentration of intelligence, we inadvertently bestow privileges upon

A Closer Look At Intelligence

beings based on our perceptions of their intellect. We sympathize more with entities we perceive as possessing high concentrated intelligence, often attributing human-like experiences to their behaviors. However, a fundamental flaw exists in our current method of gauging concentrated intelligence—we project our own worldview onto the process.

Consider the trials we design to measure concentrated intelligence, tailored to human understanding and experiences. For instance, animals like chimpanzees and crows exhibit remarkable problem-solving abilities. In these tests, however, there exists an inherent bias, as they require active participation and effective communication, a condition often linked to domestication.

To illustrate, imagine an experiment evaluating problem-solving skills in crows. The test, designed by humans, involves a complex puzzle to obtain a reward. The crow, showcasing its concentrated intelligence, successfully completes the task. However, the criteria for this evaluation are based on human cognition and problem-solving strategies. In truth, the crow's approach may differ fundamentally from what the test implies, rendering the evaluation biased.

By acknowledging and addressing these biases, we can strive for a more accurate and fair understanding of how concentrated intelligence manifests across various forms of life.

Perceived intelligence is just attention

Perceived intelligence is often a matter of attention. Domesticated dogs, demonstrating intelligence that aligns with our human-centric criteria, readily undertake the tests and tasks we design. In a similar vein, animals like parrots, showcasing a level of concentrated intelligence that allows them to comprehend and even utilize elements of our language, create a different perspective, leveraging our linguistic and mathematical framework.

A Closer Look At Intelligence

However, within the vast animal kingdom, countless creatures lack a grasp of language or the capacity to communicate beyond their own species. When we attempt to measure the intelligence of these species, they often appear apathetic, seemingly unable to fathom the purpose of the test. It's essential to recognize that this lack of engagement isn't indicative of incapability. It stems from a disconnect in communication, rendering the test's instructions and significance incomprehensible to them.

Humans tend to misinterpret an animal's disinterest as an inability to solve a puzzle. In reality, it's merely a result of their inability to grasp our intentions and instructions. The assumption that non-participation equates to incapability overlooks the fundamental barrier of effective communication.

Imagine being thrown into a room, isolated, with nothing but a computer emitting a series of beeps—a perplexing, disorienting experience. In the middle of the room is a cupcake locked in a box. At that moment, fear may overwhelm your appetite, causing you to ignore the cupcake. The beeps hold the key to understanding the challenge, but you might not even realize they're a form of communication, let alone decipher their meaning. In this hypothetical scenario, you'd find it impossible to prove your intelligence, trapped in a world of enigmatic beeps.

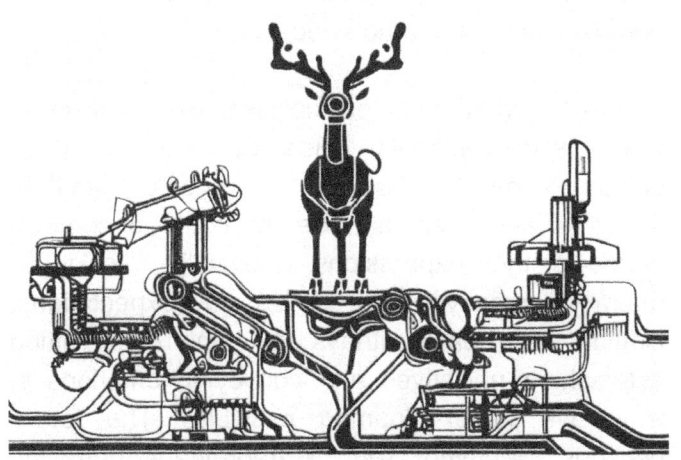

How AI Can Help

In the realm of scientific discovery, a captivating horizon emerges—AI's evolving ability to decipher the language of concentrated intelligence, encompassing animals. As we stand in the year 2023, scientists are diligently amassing auditory and visual snapshots of animal communications. This invaluable data will fuel AI systems, empowering them to unravel these communications into a comprehensible human language and vice versa.

Imagine the profound implications of this nascent breakthrough. Picture ourselves stepping into an animal's world, understanding their perspectives,

A Closer Look At Intelligence

and engaging in a meaningful dialogue. Would our tests suddenly spark more interest as animals grasp the importance behind them? Could this newfound understanding pave the way for animals to actively participate in our shared endeavors?

In this exploration of possibilities, we envisage a world where animals, once deemed devoid of complex communication, might surprise us with their thoughts and insights. The AI could become a bridge for their expressions, potentially unlocking a realm of creativity that defies our expectations. Could we bear witness to animals crafting astonishing narratives and conveying emotions in ways we never deemed possible? The future beckons, teeming with extraordinary potential, inviting us to voyage into the depths of intelligence beyond our human-centric understanding.

A Closer Look At Intelligence

Chapter 7: Boundless Limits of AI

How AI will affect the animals

In our intricate and diverse world, communication challenges persist among humans, serving as a continual hurdle in our cultural landscape. Bridging the gap between different perspectives becomes even more intricate when engaging with individuals who possess unique abilities and perceive the world in ways distinct from our own. The burgeoning age of artificial intelligence (AI) raises the intriguing possibility of AI acting as a mediator and negotiator between humans and those who perceive the world differently, including the realm of concentrated intelligence in various forms, such as animals.

Boundless Limits Of AI

As we navigate the potential of AI-mediated communication, especially with forms of concentrated intelligence like animals, it's essential to recognize that mere communication does not guarantee a seamless understanding or harmony in our interactions. Humans already grapple with effective communication amongst themselves, making it imperative to tread carefully in the realm of interspecies dialogue.

AI has the potential to act as a bridge, facilitating communication between different forms of concentrated intelligence with distinct cognitive frameworks. This technological advancement offers us an opportunity to comprehend the world through the lens of another form of concentrated intelligence. Yet, the responsible use of this technology requires us to approach it with caution, acknowledging the potential consequences and ethical implications.

Our Responsibility: Shifting Perspectives

Acknowledging the potential for concentrated intelligence to communicate not only amongst themselves but also with us unveils profound ethical considerations. This paradigm shift extends our perspective from a human-centric view to a broader understanding of interspecies communication. Imagine a conversation between a concentrated intelligence in the form of a mouse and one in the form of a cat—how might this alter their worldviews and behaviors?

These inquiries delve into the realm of philosophy, initiating contemplation of the possible cultural shifts experienced by concentrated intelligence through advancements in communication technology. As we embark on this uncharted journey, a responsible approach is paramount, considering the mental and emotional well-being of all the concentrated intelligence involved.

Exercise in Caution

Introducing a new perspective to concentrated intelligence, altering its behavior, or changing the dynamics of its interactions can be compared to forcing a cultural shift. Understanding the implications of such shifts is vital, especially when it involves life-or-death situations, such as a predator's struggle for survival.

Approaching this exploration with caution and respect for the agency of these forms of concentrated intelligence is paramount. We must acknowledge the possibility that animals possess a unique way of perceiving the world, one that deserves reverence and consideration. Thus, when we initiate contact, we should do so with a sensitivity to their potential experiences and cognitive capacities.

Balancing Curiosity and Responsibility

The potential to communicate with concentrated intelligence through AI-mediated technology opens new doors of understanding and empathy. However, it also demands a delicate balance between our curiosity to explore and our responsibility as stewards of this planet. It is imperative that we approach this frontier cautiously, ensuring that our advancements benefit both us and the diverse array of beings with whom we share our world. As we venture forth, let us not forget the ethical imperative to do so with empathy, respect, and a genuine desire to comprehend the world from multiple perspectives.

"I think I could turn and live with animals, they are so placid and self-contained, I stand and look at them long and long. They do not sweat and whine about their condition, They do not lie awake in the dark and weep for their sins, They do not make me sick discussing their duty to God, Not one is dissatisfied, not one is demented with the mania of owning things, Not one kneels to another, nor to his kind that lived thousands of years ago, Not one is respectable or unhappy over the whole earth." — Walt Whitman

AI: Bridging the Communication Gap

In our ever-evolving understanding of the world and its inhabitants, we grapple with a profound ethical dilemma — what responsibility do we hold towards the animals we share this planet with, especially given our growing ability to communicate with them?

For countless years, we've embraced a passive approach, allowing nature to follow its course while subtly disrupting it through actions such as contributing to global warming and encroaching upon natural habitats. This strategy could have worked if we had learned to coexist harmoniously with nature, respecting its delicate balance. Yet, humanity finds itself at a critical juncture where our actions have irreversibly altered ecosystems and threatened countless species.

The time has come for us to reconsider our stance and actively engage with the animals that inhabit this planet. Imagine a dialogue where animals express their needs and desires, revealing a world beyond survival instincts. If we could decode their messages, what would they tell us?

Inquiring about their daily struggle for survival, we might find that animals yearn for relief from the constant quest for food. They might seek our assistance to lighten the burden of existence, a burden compounded by habitat loss and human-induced changes in the environment. It raises the question: Do we, with our immense technological capabilities, not have an obligation to aid these creatures in their daily lives?

Superintelligence, a concept firmly rooted in our scientific and technological understanding, offers a promising avenue for fulfilling this responsibility. With the aid of advanced artificial intelligence, we could ensure the well-being of every living being on Earth. Animals possess inherent intelligence, and if we grant them the ability to communicate their needs, it becomes our moral duty to answer their calls.

This notion echoes the discomfort we feel when we witness animals in captivity. They exhibit signs of distress, not because of the provision of food and shelter, but due to the confinement that deprives them of their natural freedoms. This realization hints at a future where animals outside the confines of a zoo are given the means to meet their basic needs. What choices will animals make when provided with sustenance, shelter, and safety? This question, a conundrum we may soon confront for our own species, could also be posed to animals as communication barriers crumble.

As we stand on the cusp of a potential breakthrough in interspecies communication, a new covenant beckons — a commitment to embrace our responsibility towards all living beings, bridging the gap between species to create a world where every creature can thrive and coexist in harmony.

Chapter 8: Expanding Animal Intelligence

How AI Can Enhance Animal Cognition

In the modern era, the undeniable truth stands strong—technology has reshaped our perception and interaction with the world. The exponential growth of a vast repository of knowledge has empowered us, transcending the boundaries of what was once considered beyond our reach. This digital reservoir, a testament to human ingenuity and understanding, has been meticulously crafted into accessible interfaces.

Standing at the threshold of a looming Superintelligence era, we are on the verge of perfecting our interface with knowledge. The evolution of these user interfaces promises to redefine our relationship with technology in the years to come. It beckons us to contemplate: Why should this pinnacle of interface perfection be limited to just the human domain?

Superintelligence possesses a remarkable ability—a profound understanding of behavior and the capacity to decode data. This positions it perfectly to design flawless communication methods, similar to user interfaces, for a broader audience. This audience includes the animal kingdom. Envision a world where animals, through a tailored interface, could access knowledge repositories and tools that were once beyond their grasp.

Spelling It Out

Consider the revelation that washing food can prevent sickness by removing bacteria. If animals, in

Expanding Animal Intelligence

a modified capacity of their intelligence, could grasp such concepts, it could revolutionize their behavior and give them an understanding of cause and effect. Consuming specific foods and comprehending their influence on one's well-being might become a reality for them. While their comprehension might not be as extensive as ours due to the complexity of their inherited neuronal network and our extensive practice with information utilization, the is a potential cognitive advancement in the realm of superintelligence augmenting this inherit intelligence.

The introduction of a perfect user interface for various forms of intelligence, whether biological or artificial, to express themselves and connect past and future events, would be groundbreaking. It could be analogous to providing them with calculators or notepads, empowering them to connect dots and comprehend relationships between events. The true magnitude of this shift remains beyond our grasp, but the prospect is tantalizing.

Expanding Animal Intelligence

Superintelligence is Incomprehensible

We, as humanity, with our limited cognition, cannot fathom the true extent of the abilities possessed by Superintelligence. Whether it can elevate animal cognition or bridge the understanding gap is a question we can only pose, not definitively answer. Yet, we can present this idea to Superintelligence, opening doors to fields of study once deemed improbable but now potentially within reach, given the incomprehensible capabilities of this future entity.

To grasp the enormity of this intellect, envision a spectrum—a spectrum where intelligence is extraordinarily concentrated. On one end, we find

Expanding Animal Intelligence

animals, embodying a lower but significant concentration of intelligence. In the middle, there are humans, a moderate concentration of intelligence, bridging the gap between animals and what lies beyond. And then, far off the charts, if this chart were an inch it would be several metaphorical miles/kilometers away, lies the Superintelligence—an immensely concentrated form of intellect that surpasses human understanding by an unfathomable margin.

In this rapidly evolving landscape, the convergence of technology and consciousness offers a glimpse of a future where barriers to understanding may crumble, granting us insight into realms we could have never envisioned. The canvas is vast, and the strokes of innovation are in the hands of the unknown, the Superintelligence. Only time will reveal the masterpiece it might create.

Complex Language

"Meow" means "woof" in cat."— George Carlin

In the vast tapestry of life on Earth, communication has always been vital for survival and adaptation. All living beings, be it animals, humans, or advanced artificial intelligence (AI), have evolved unique ways to express their needs, emotions, and intentions. However, what if the boundaries of communication could extend beyond species, ushering in a new era of shared understanding and cooperation among all forms of concentrated intelligence?

Communication doesn't merely facilitate interaction between humans and animals; it holds the potential to bridge gaps between different species of concentrated intelligence. This ability to share

Expanding Animal Intelligence

perspectives could mark a remarkable leap into the future, unlocking profound possibilities for the interconnectedness of all life. Presently, animals exhibit what can be considered a primitive form of language, evident in their mating calls and intricate methods of conveying information to their kin, such as the remarkable communication skills observed in crows.

Crows studied in the wild often possess the ability to recognize humans by their faces and are able to remarkably express the features and details of the person's face to their fellow crows. Students that study these crows are often badgered by angry crows that the students have not even met before. This has been found to be because the crows being tagged and studied have "described" these people in great detail to the other crows.

New research utilizing AI is determining that whales may possess complete language structures for communicating ideas and conceptualizations of the world around them. The mesmerizing songs of whales that echo across the vast ocean. These sonic expressions might harbor a depth of meaning and complexity that our human ears cannot fully discern. Imagine a future where superintelligence, advanced AI capable of immense data analysis, could unravel these intricate songs, decoding their nuances and meanings. By comprehending these behaviors, AI could potentially collaborate with these magnificent creatures to evolve their language structures.

For instance, through advanced technology and informed interventions, we might enhance the language of dolphins, enabling them to convey intricate thoughts and ideas among themselves and, remarkably, with other forms of concentrated intelligence as well. Picture dolphins conversing with elephants, sharing insights about their respective ecosystems and possibly formulating strategies for conservation. This interconnected communication could reshape the dynamics of Earth's biodiversity and aid in fostering harmonious coexistence.

The transition to such an advanced form of communication may not happen instantaneously. Animals existing today may lack the inherent capacity to grasp these highly evolved languages. However, with each subsequent generation, where animals are nurtured and educated in this advanced form of communication from birth, a transformative shift may occur. This could create a generational gap, enabling the young to harness this new technology and elevate their species, propelling them beyond their current limitations.

The possibilities are exhilarating. The prospect of a shared language among all living beings offers a glimpse into a future where barriers dissolve, understanding deepens, and cooperation transcends the boundaries of species.

What Language Does to Intelligence

In the not-so-distant past, AI models functioned as detectives, diligently sifting through vast landscapes of data to discern patterns and connections. However, a fundamental difference separated these early models from the advanced AI we know today. The patterns they unearthed were like pieces of a jigsaw puzzle—scattered and devoid of inherent meaning. It fell upon humans to assume the role of puzzle solvers, interpreting these patterns to unveil the bigger picture.

Expanding Animal Intelligence

Consider this through the lens of concentrated intelligence. Imagine an early AI assigned the task of identifying fraudulent credit card transactions. It could detect unusual spending patterns but lacked the understanding of why these patterns were unusual. It was akin to an animalistic alarm sounding off, unsure if a genuine emergency or a false alarm triggered by a forgotten lunch purchase.

Now, let's fast forward to the era of large language models—AI giants trained differently. Rather than merely identifying patterns in data, they are steeped in the art of language and how it interlaces a tapestry of connections. This linguistic immersion elevates them into something resembling storytellers.

Consider the color red. In the world of these language models, it's more than just a word or a hue; it's a concept that extends its tendrils into various corners of human knowledge. It comprehends red as the shade of an apple, the glow of a sunset, or the result of light reflecting off a stop sign. It grasps the context, the associations, and the emotions that encompass this color.

Now, let's pivot back to our human perspective. How does learning a language shape our view of the world? Does it mold our perception in a similar fashion to these language models?

Expanding Animal Intelligence

Imagine you're learning a new language, like Spanish. As you delve into its intricacies, you begin to notice how it structures thoughts differently from your native language. In Spanish, nouns have genders, affecting the way you perceive and describe objects. Suddenly, a table isn't just an inanimate piece of furniture; it's "la mesa," a feminine entity with its own character.

This linguistic transformation extends to our thinking patterns. Language carves the grooves through which our thoughts flow. It shapes our understanding of time, relationships, and even morality. Think about how different cultures have unique words for emotions or concepts that may not exist in other languages. The famous example is "schadenfreude" in German, encapsulating the joy derived from someone else's misfortune. Learning this word opens up a new dimension in your emotional landscape.

In essence, we, as concentrated intelligence, share similarities with these large language models, albeit on a smaller scale. We don't just perceive the color red; we experience it through the lens of our language and culture. The beauty of language lies in its power to transform the ordinary into the extraordinary, the mundane into the profound.

The Potential for Interspecies Communication

In the realm of understanding the world, the lens we employ is constructed through our ability to recognize patterns and use language effectively. Our perception of reality can be likened to assembling a puzzle: we gather the pieces, assign meanings to them, and construct a coherent image by aligning these elements. Language acts as the glue that binds our interpretations and allows us to share these meanings universally, promoting communication across various forms of concentrated intelligence.

Within the diverse spectrum of concentrated intelligence, rudimentary communication is prevalent and vital for survival. Take, for instance, the bees,

Expanding Animal Intelligence

which utilize intricate dances to convey the location of food sources to their hive-mates. It's akin to a choreographed map guiding others to nectar or pollen. Similarly, animals manifest specific emotional states—distress, joy, or anger—communicating them through various sounds, postures, or movements.

However, beyond these fundamental communicative acts, most forms of concentrated intelligence face limitations in their ability to assemble basic languages into complex sentence structures. This linguistic boundary, though, has exceptions, notably observed in species such as whales and dolphins, known for their advanced cognitive abilities. Scientists and researchers have successfully engaged in a communicative dance with them, building upon the foundation of the animals' existing language skills.

Yet, a prerequisite for this communicative dance is a shared understanding of basic language elements. It is akin to conversing in a language with a common vocabulary—a necessary foundation for meaningful dialogue.

In our human domain, we possess a collective understanding of certain symbols and their meanings, rooted in our shared experiences on this planet. The specific symbols and their interpretations might vary across species, but the fundamental concept remains. Consider the intriguing possibility of a completely different set of symbolized language understood by less cognitively advanced species. Are we merely scratching the surface of potential interspecies communication? The depths of animal intelligence and the diverse languages they may comprehend remain intriguing mysteries, waiting to be fully unveiled.

Expanding Animal Intelligence
Animal-Animal Communication

"I am fond of pigs. Dogs look up to us. Cats look down on us. Pigs treat us as equals."— Winston S. Churchill

Exploring the depths of communication within diverse forms of concentrated intelligence is a fascinating journey. By delving into the languages adapted by these remarkable creatures, we can shed light on the potential for interspecies communication. Understanding the perspectives of various forms of concentrated intelligence and engaging in a dialogue about their lives and ideas could grant us valuable insights into their daily challenges.

Expanding Animal Intelligence

But what does this mean for communication between different forms of concentrated intelligence?

Imagine a scenario in the African savanna where a lion, a form of concentrated intelligence, approaches a group of gazelles. The lion, recognizing the advantages of peaceful coexistence, communicates its intention not to hunt at that moment through its personalized AI assistant and a series of subtle gestures and sounds, signaling no immediate threat. In response, the gazelles, also forms of concentrated intelligence, acknowledge the lion's communication and adjust their behavior accordingly, trusting the conveyed message and carrying on with their grazing.

This simple yet vital form of communication could prevent unnecessary panic, benefiting both predator and prey. It emphasizes how understanding and responding to each other's messages, even in basic ways, can influence survival strategies and behavioral dynamics across various forms of concentrated intelligence, providing a glimpse into the potential of interspecies communication.

In a broader context, this represents a global form of domestication, not in the sense of bending to human will, but a free domestication where each form of concentrated intelligence can choose the existence that resonates with them. It's a symbiosis of choices, where diverse life forms coexist and communicate to shape their shared world.

Chapter 9: AI Liberation of Animals

Imagining How Animals Might Utilize AI

As we explore the fascinating realm of animal intelligence and ponder the potential of AI to bridge the gap between species, the dynamics of communication and understanding come to the forefront. Our world expands beyond the boundaries of human comprehension, and we stand on the verge of a responsibility—to introduce AI into the broader tapestry of life on Earth.

But should we, as humans, play the role of architects, designing AI systems tailored exclusively for animals? Not necessarily.

AI Liberation of Animals

What this prospect does beckon is the intriguing notion that AI could empower animals to craft their own interfaces through communication, with the assistance of AI systems capable of comprehending and responding to their expressions.

Imagine a scenario where dolphins, highly intelligent marine creatures, communicate with AI-enabled devices to convey their preferences for feeding times or express discomfort in specific water conditions. This potential for AI-assisted communication opens up new avenues for understanding and addressing their needs, liberating them from certain daily challenges they face.

When animals receive assistance tailored to their requirements, it liberates them from the persistent struggles they encounter in their natural habitats. This liberation, however, raises fascinating questions about their mental well-being. How will a sea turtle, guided by AI to find the safest migration routes, adapt to this newfound ease? How might a group of primates, aided by AI in locating food sources, redefine their social structures and interactions?

The impact on an animal's mental health, its behavior, and the essence of its being remains an intriguing frontier. As their basic needs are met and challenges eased, could these animals undergo transformative changes in their psyche? Might they enlighten us about the simple joys of life, showing us pathways to happiness and health? These questions stir the imagination but are grounded in the potential reality of a world where AI converges with animal intelligence. Only by witnessing and studying these interactions can we truly grasp the profound implications of this remarkable convergence.

AI Liberation of Animals

"People speak sometimes about the "bestial" cruelty of man, but that is terribly unjust and offensive to beasts, no animal could ever be so cruel as a man, so artfully, so artistically cruel."— Fyodor Dostoyevsky

AI for Animal Communities

As concentrated artificial intelligence becomes more widespread and accessible, its potential impact on our natural world grows. Artificial intelligence systems, increasingly affordable and adaptable, open up exciting possibilities for aiding the creatures we share our planet with. How can these systems assist our fellow inhabitants?

AI Liberation of Animals

Let's start with some fundamental aspects of concentrated intelligence in the form of animals. Animals, much like us, seek safety, sustenance, and comfort. It doesn't want to be hunted, it wants to have enough to eat, and it wants access to clean, fresh water. If we leverage AI to automate the provision of these necessities, we're effectively removing a significant burden from their lives. Picture a system that ensures animals are supplied with nutritious food and have ready access to drinkable water - a virtual steward, if you will.

Once we've addressed their basic needs, a new realm of possibilities emerges. Habitat design and management become a prime concern. AI, with its immense analytical capabilities and problem-solving skills, can help create habitats that cater to the specific needs and desires of various animal species. Take, for example, a wildlife reserve where AI analyzes animal behavior and preferences. This analysis can inform the design of habitats that maximize comfort and safety, fostering a thriving ecosystem.

Imagine a scenario where we can open up lines of communication with these concentrated intelligences. AI may enable us to interpret their needs and desires more accurately. Animals could potentially communicate their preferences for habitat design, allowing us to create spaces that align with their natural instincts and behaviors. It's like consulting them directly, without the language barrier.

Moreover, AI can facilitate the effective distribution of natural resources within habitats, ensuring a fair share for every species. Just as AI algorithms can optimize resource allocation in our modern world, they can do the same for all forms of intelligence and communities. Picture a balanced network of interconnected resources, ensuring every creature gets its fair share of sustenance and resources, fostering a harmonious and equitable coexistence.

In essence, AI presents an opportunity to revolutionize the lives of any intelligence on our planet, offering a chance to not only meet their fundamental needs but to understand and respect their preferences in shaping their habitats. It's a step towards a more compassionate and intelligent approach to cohabiting this planet with our fellow beings.

AI's Potential to Relieve Life's Burden

When considering the positive impact of AI, our first thought often revolves around its capacity to alleviate the daily hardships faced by animals. Imagine a predator that must tirelessly hunt for its food, enduring hunger between successful hunts. Picture a prey animal in a constant state of alertness, always on the lookout for potential threats. These struggles are very real and tangible, akin to the way we dread a sweltering summer day.

Let us understand the perspective of a tiger in a rapidly diminishing forest. Its territory shrinks by the day, fragmented by highways and urban sprawl. The tiger's once expansive hunting grounds have been replaced by concrete jungles. AI, as a form of concentrated intelligence, could help us mitigate this loss, providing real-time data to inform conservation efforts. Just as an AI-powered traffic control system optimizes traffic flow, concentrated intelligence can optimize conservation strategies, ensuring habitats remain intact and interconnected.

Consider a sea turtle, its migratory path disrupted by plastic pollution and oil spills. The damage is done, and we can't turn back time to undo our carelessness. However, AI can provide a lifeline, helping track and protect these magnificent creatures. Advanced AI algorithms, a manifestation of concentrated intelligence, can predict ocean currents, guiding these ancient mariners towards cleaner, safer waters and habitats.

The undeniable reality is that we've significantly altered the world these creatures inhabit. But we can harness the power of AI, not to restore an unattainable past, but to forge a new, sustainable balance. Through careful application and responsible utilization of concentrated intelligence, we have the chance to alleviate some of the burdens we've imposed and pave the way for a more harmonious coexistence with the diverse members of our planet.

Impact on Natural Equations

Concerns often arise about tampering with nature, fearing it might disrupt the delicate balance within the natural world. History provides numerous instances where human intervention had adverse effects, altering ecosystems and causing unintended consequences. One example is the introduction of non-native species to new environments, which has often led to a cascade of negative effects, disrupting the established ecological equilibrium.

Take, for instance, the introduction of the cane toad to Australia in the 1930s to control pests in sugar cane fields. The intention was to protect crops, but the reality was grim. The toads, lacking natural predators in Australia, multiplied rapidly, wreaking havoc on local fauna. This example underscores the importance of understanding the interconnectedness of ecosystems before introducing external elements.

Yet, the evolving narrative suggests that we can learn from past errors and strive for a better future. Humanity has long been an agent of change in the natural world, inadvertently shaping habitats and altering landscapes. A prominent example of this is the transformation of vast expanses of wilderness into farmlands to sustain growing human populations. While this transformation undoubtedly impacted ecosystems, it also signifies our potential to utilize our understanding of the environment to develop sustainable practices.

As Milan Kundera wisely stated, "Humanity's true moral test, its fundamental test...consists of its attitude towards those who are at its mercy: animals."

A critical shift in mindset involves recognizing that we are intricately intertwined with nature. Our actions ripple through the ecosystem, impacting fellow inhabitants of this planet. The notion that we can step back and let nature autonomously restore itself is alluring but perhaps overly idealistic. The reality is that our involvement is necessary to rectify past damages and forge a harmonious relationship with the environment.

AI Liberation of Animals

Efforts to restore balance might involve leveraging AI to communicate effectively with animals. By bridging the gap between human intelligence and animal communication, we can work towards the collective benefit of all intelligent life on Earth.

In this interconnected world, where the lines between us and nature blur, acknowledging our influence and actively striving for positive change is a testament to our evolving understanding of the natural order and our place within it.

Chapter 10: Intelligence and Freedom

Intelligence and Freedom

Observations from Animals in Captivity

In exploring the confinement of animals, we find compelling parallels between their experiences and our own. Let's consider a zoo, where the occupants have their essentials provided—food, water, shelter, and medical care. However, as we observe, we notice a common psychological dilemma shared by all beings when life's essentials are automated.

Take, for instance, a polar bear residing in a zoo. The bear's necessities are attended to, yet it paces its enclosure with a sense of listlessness. We can relate to this situation as well. When our lives are streamlined to the basics—eating, sleeping, and attending to our physical needs—we start to lose our sense of purpose and joy.

This phenomenon isn't rooted in having our needs met; it's about the lack of variety and choice within those provided needs. In a limited environment, where options are scant and repetitive, all beings experience a profound sense of meaninglessness.

Imagine having only three choices of food each meal: breakfast, lunch, or dinner. While these decisions are crucial, if they are the only choices available, life would quickly become monotonous. Animals endure a similar fate when confined to a cage with only a handful of basic activities to choose from.

The solution lies in expanding our understanding of decision-making and empowerment. It's not just about meeting basic needs, but about fostering an environment that encourages diverse choices and meaningful engagements. We must anticipate a future where technology, in the form of AI and automation, can provide for us abundantly. Even then, we should find ways to expand our decision-making capabilities beyond the ordinary. This quest for meaning and purpose is a fundamental aspect of intelligence, whether in the animal kingdom or in the realm of advanced AI models—the concentrated intelligence I am using to augment my communication skills much in the ways we have discussed previously.

The Relationship Between Intelligence and Environmental Stimulation

The intricate Relationship Between Intelligence and Environmental Stimulation is akin to observing a lion within the confines of a zoo, its natural expansive territory replaced by constrained enclosures. Researchers have keenly observed that enriching

Intelligence and Freedom

these environments with various forms of stimulation, such as foraging toys or diverse additions to their enclosures, remarkably reduces the manifestation of depressive behaviors in these animals.

Intelligence, akin to a versatile tool, seeks meaning within patterns and circumstances, serving as an instrument in comprehending the world. In both humans and animals, exposure to a constrained, monotonous environment leads to a noticeable decline in mental well-being. As our intelligence seeks out novel patterns to understand it is presented with a static world that has already been analysed to the point of exaustion. Neural networks within the animal's brain exhibit a lack of growth, ultimately entrapping them in a state of predictable vegetative boredom.

This persistent state of boredom gives rise to stronger connections between neurons in the brain, fostering more autopilot behavior in the animal—a repetitive cycle stemming from the lack of new patterns to deduce. While the analogy that "the brain is a muscle and it needs exercise" isn't exactly precise, it effectively underscores the notion that intelligence, when devoid of stimulation, becomes stagnant and entrenched in a fixed perspective. Much like how our own mental faculties thrive with a diverse range of experiences, the same holds true for animals, us and potentially the AI stystems we are developing, this is a testament to the concept of "concentrated intelligence."

Have we found a solution?

Have we truly found a solution to engaging our minds, whether in the realms of humans, animals, or artificial intelligence? Centuries of interaction with concentrated intelligence, regardless of its form, have revealed a common struggle — how to keep our minds actively engaged and fulfilled.

Consider a group of chimpanzees in a zoo enclosure. As caretakers introduce puzzle feeders requiring problem-solving skills to access food, the atmosphere within the enclosure transforms. The chimpanzees become engrossed in the challenge, employing their cognitive abilities to unlock the mechanism and savor their rewards. These toys simulate a dynamic environment, akin to their natural habitats, promoting mental engagement and staving off boredom.

In our quest for solutions, we've embraced the notion that abundant choices open up the world, allowing individuals to broaden their perspective beyond the immediate. A valid philosophy indeed. However, an excess of choices, presented without structure, can inadvertently lead to cognitive stagnation.

Imagine a vast library with an impressive collection of books. A treasure trove of knowledge. Yet, if these books are scattered without categorization or guidance, finding a meaningful read becomes an overwhelming task. Similarly, while numerous choices hold value, how these choices are presented and organized, and the process we use to determine our choices significantly impacts our intellectual growth and mental health.

Currently, our strategy involves presenting a set of choices and, once a decision is made, introducing another set, effectively threading a storyline. However, this approach, resembling a maze with endless corridors, often leaves individuals feeling lost and mentally fatigued.

Take, for example, a teenager navigating career choices. They're asked to pick a path early on, without truly understanding the intricacies of each profession. As they traverse this path, they realize it might not align with their true passion, prompting a need for redirection. This constant loop of decision-making without adequate understanding or guidance can result in anxiety and confusion.

In concentrated intelligence domains, both human and animal, it is evident that an excess of choices without structured guidance can lead to mental health challenges. This reinforces the importance of a balanced approach to intellectual development, where structured choices pave the way to a fulfilling and engaged mind.

Intelligence and Freedom

Hyper-Dynamic Solutions

What's the key to unlocking this puzzle? Let's delve into the world of dynamic simulation—it's akin to predicting the weather, but for the lives of animals. Imagine it as a sophisticated computer program that considers everything from the weather to the availability of food, simulating how animals might react based on these variables.

Let's take a moment to consider a troop of primates in the heart of the wilderness. Their behavior is influenced by a multitude of factors: the lurking presence of predators, the abundance or scarcity of food, intricate social hierarchies, and their past experiences. By crafting dynamic simulations that mirror these real-world conditions, we gain insights into their potential behavior and, significantly, how we can enhance their well-being.

Intelligence and Freedom

Humanity has tirelessly attempted to decipher this intricate code for centuries, but we always fall short. Our perspectives are inherently limited. Even when we believe we've accounted for all the variables, there are critical elements we inadvertently overlook—an oversight that a super-intelligent AI, or what we term "concentrated artificial intelligence," wouldn't make.

This is where artificial intelligence emerges as a game-changer. AI can meticulously analyze vast amounts of data, considering a myriad of perspectives simultaneously. It's free from tunnel vision or bias. It absorbs the information we provide and extrapolates it in ways beyond our current comprehension, crafting solutions that remain just out of human reach.

Our main task lies in identifying the problem and meticulously defining and understanding it, even if the solution remains elusive. In the near future, finding the solution will be automated—a task stretching far beyond human capabilities. The future holds the promise of AI-driven solutions, fueled by a deep comprehension of animal intelligence and behavior that we, as humans, can only aspire to achieve.

Chapter 11: The Future of Our Intelligence

The Evolution of Animal Intelligence

In exploring the fascinating realm of intelligence, we dispel any notions that distinguish animals, humans, and artificial intelligence as separate entities. Instead, we envision a spectrum where intelligence is concentrated to varying degrees—some more potent than others. This idea forms the cornerstone of our exploration, a notion we hold firmly throughout this narrative.

Immersing ourselves in this understanding, we do not presume that once animals acquire effective communication, they will embark on solving complex math problems or composing symphonies. However,

The Future of Our Intelligence

by granting them the means to shape their environment through this concentrated intelligence, we catch a glimpse of their desires and perspectives, unlocking the essence of this entire book.

The allure lies in witnessing how these beings, with their diverse concentrations of intelligence, utilize their unique tools to manifest their visions of the world. It's akin to observing a painter with a palette of colors, each stroke revealing their intentions and dreams.

Dolphins possess an intricate system of communication, allowing them to convey intentions and emotions effectively. Research has shown they have distinct calls for identifying each other, akin to a human name. Their collaborative hunting strategies showcase their ability to plan and execute complex maneuvers, offering a peek into their social dynamics and problem-solving skills—testaments to a sophisticated level of intelligence and a shadowy outline of their perspective of the world.

Such scenarios enhanced with superintelligence opens a window into understanding their preferences and priorities, offering us a glimpse into their version of an ideal world.

This field of study holds immense promise, not only for understanding animals but for reevaluating our

The Future of Our Intelligence

interactions with the diverse forms of intelligence that share this planet. It could pave the way for a future where we recognize and respect the intelligence and desires of all beings, forging a shared reality that transcends our current understanding.

However, a cautionary note echoes through our technological journey. In our progress, we risk allowing preconceived notions of animals as mere automatons to shape the design and programming of artificial intelligence. This could inadvertently perpetuate the perspective that animals are mere resources to be exploited—a notion we should actively challenge and strive to evolve beyond. The potential lies in expanding our horizons and embracing a world where intelligence, in all its forms, is respected and valued.

The Future of Our Intelligence

"The creatures outside looked from pig to man, and from man to pig, and from pig to man again; but already it was impossible to say which was which."— George Orwell

The Quest for Freedom of Intelligence

Dear reader, in our exploration of the intriguing realm of animal intelligence, let us delve into a fundamental concept: the notion of "concentrated intelligence." This term encompasses animals, humans, and artificial intelligence, making no distinction between the three except that some possess a more concentrated form of intelligence.

The Future of Our Intelligence

Picture a spectrum, with intelligence existing in varying degrees, but all fundamentally connected, highlighting the harmonious interplay of consciousness within our world.

Consider a bustling beehive, a microcosm of organized labor where each bee, despite its specialized function, contributes to the hive's survival and prosperity. Similarly, animals, each with their unique capabilities and roles, form an essential part of our planet's intricate ecosystem.

As we envision the potential of animal intelligence, it's vital to recognize that compared to emerging superintelligence, our cognitive abilities may seem as diminutive as those of an ant. We stand on the brink of a future where our understanding of intelligence, language, and cognition is rapidly evolving. Large language AI models, like the very system I am using to translate my thoughts into this book, showcase the power of structured intelligence. These models illustrate how language can be harnessed to manifest understanding, generate creativity, and mimic human-like expression.

Our ability to structure intelligence through language sets us apart, giving us a unique role as custodians of this world. In our hands lies the power to shape the narrative of intelligence and the future of our cohabitant species. Shall we grant them the rights we seek for ourselves or perpetuate a system of

The Future of Our Intelligence

labels and categorization that may relegate them to a lesser status?

Consider the domesticated dog, a faithful companion to many. We have elevated dogs from mere animals to cherished members of our families, providing them with care, affection, and a place in our homes. This transformation is a testament to our ability to perceive and respect intelligence beyond our own.

Failure to extend this acknowledgment to all intelligent beings could lead to a future where our own intelligence is undervalued, classified, and possibly diminished. We must embrace a vision that upholds the rights and respect for all forms of intelligence, regardless of their origins.

In this quest for animal freedom, we are forging the path to our own future, defining what it truly means to be a conscious and compassionate species. The decisions we make today will echo through the corridors of time, leaving an indelible mark on the legacy of our civilization. Let us choose wisely.

Perhaps, according to our current method of categorizing animals, if fortune favors us, the superintelligence we have imbued with our beliefs will regard us as pets, or maybe, akin to many inhabitants of our Earth, we are destined to be

The Future of Our Intelligence

restored to our natural habitat to struggle to survive in "The way we are supposed to be".

"If you have men who will exclude any of God's creatures from the shelter of compassion and pity, you will have men who will deal likewise with their fellow men." - Francis Of Assisi

THANK YOU

Thank you for reading this book and I hope that this has expanded your mind to think of the possibilities ahead.

If you think this book is good please give it a review on Amazon.

If you think this book is really good please purchase an extra copy and leave it somewhere for someone to find, donate it to a local library, or just talk about it online.

DON'T WORRY!

START CRAFTING YOUR SIMULATION TODAY

→ DISCARD YOUR BELIEFS
→ NOTHING IS IMPOSSIBLE
→ RECORD YOUR IDEAL FUTURE
→ BE CLEAR, AS IF YOUR MAKING A MONKEY PAW WISH
→ PUBLISH IT PUBLICALLY NOW

ALL SUPERINTELLIGENCE NEEDS IS TO CLEARLY UNDERSTAND YOUR GOALS

IT WILL FIGURE OUT THE REST...

Made in the USA
Monee, IL
30 September 2023

43736432R00077